The Physiology and Cosmology of Consciousness:

The Pribram-Bohm Hypothesis

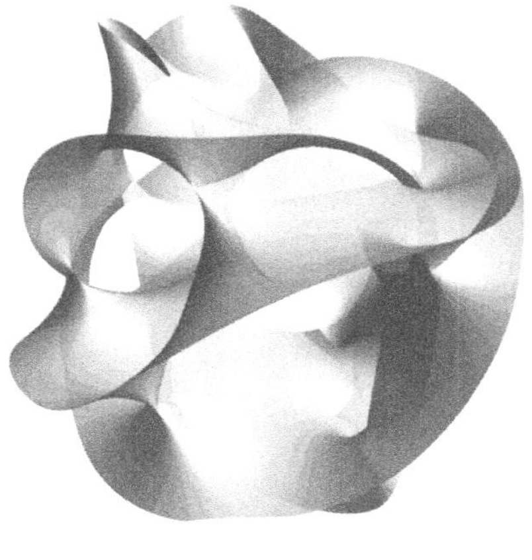

Shelli Renée Joye, PhD

B.S. Electrical Engineering
M.A. Indian Philosophy
Ph.D. Philosophy, Cosmology, and Consciousness

Published by the Viola Institute
Viola, California

2022

Dedicated to my teachers, friends, and mentors,
John Lilly, Dean Radin, Allan Combs,
Brian Swimme, Bruno Barnhart, Robert McDermott,
and Chögyam Trungpa Rinpoche.

Published in the United States of America

Some of the material in this book has appeared in the author's previous books: *Sub-Quantum Consciousness, The Electromagnetic Brain, Tuning the Mind,* and *The Little Book of Consciousness.*

Printed in the United States of America

The Physiology and Cosmology of Consciousness:
The Pribram-Bohm Hypothesis

ISBN: 978-1-950761-13-5

CONTENTS

KEY TERMS AND DEFINITIONS

Explicate Order: One of the two primary dimensional orders in the universe, as posited by David Bohm. The explicate order encompasses spacetime and all manifestations within spacetime. The other dimensional order Bohm termed the "Implicate Order."

Holosphere: A spherical boundary that has the diameter of one Planck length (3×10^{-35} m).

Holoplenum: A continuous plenum of Planck length holospheres at the bottom of space.

Holoflux: Energy-phase-information flux within Bohm's implicate order; it corresponds to electromagnetic flux in spacetime.

Implicate Order: One of two primary dimensional orders in the universe, as posited by David Bohm. The implicate order is a nonlocal, nontemporal domain that can be found below the Planck length. In holoflux theory it is equivalent to the frequency-phase domain in communications engineering. Bohm felt that the implicate is "somehow 'deeper' and more fundamental than the explicate" (Peat, 1997, p. 262).

Planck length: The smallest possible length in space, determined by Max Planck from the gravitational constant and the speed of light. The value of the Planck length is 3×10^{-35} m.

Planck time: The smallest possible unit of time, determined by Max Planck from the gravitational constant and the speed of light. The value of Planck time is 5×10^{-44}

Introduction

Perhaps the most widely debated issue in consciousness studies can be found encapsulated in the phrase "the hard problem of consciousness," first articulated in an essay by David Chalmers (1995) in which he discusses consciousness as experience. A product of our digital age, Chalmers understandably conflates experience with information processing, but questionably suggests that it is information processing itself that gives rise to experience:

> *The really hard problem of consciousness is the problem of experience. When we think and perceive, there is a whir of information processing, but there is also a subjective aspect. . . . Why should physical processing give rise to a rich inner life at all? It seems objectively unreasonable that it should, and yet it does. If any problem qualifies as the problem of consciousness, it is this one. (p. 5)*

Chalmers (1995) here places the cart before the horse if he really means to say that consciousness arises from information processing. That would make consciousness an epiphenomenon, something that arises from what he calls the "whir of information processing." What he seems to be addressing is the hard problem of mind, rather than the hard problem of consciousness. More recently, however, Chalmers (2010) has suggested that consciousness may not be derivative after all, that it may be seen as "fundamental" in the same sense as space and time are regarded by physicists as "fundamental."

Like epiphenomenalism, other assumptions restrict the range of contemporary approaches to consciousness research, and may accordingly be similarly misleading, for example (a) the assumption that the word consciousness is limited specifically to "human consciousness," and (b) the assumption that consciousness is a phenomenon exclusive to the spacetime continuum.

Understandably, such assumptions have arisen due to a widespread fascination with hardware, and in particular the hardware of the brain, fostered by an approach to science which limits itself to measurements which can be observed in space–time. Thus epiphenomenalism has become mainstream, reinforced by such statements as the following, found in Gerald Edelman and Giulio Tononi's (2000) conclusion to their book, A Universe of Consciousness, where they observe:

> *Consciousness while special, arose as a result of evolutionary innovations in the morphology of the brain and body. The mind arises from the body and its development; it is embodied and therefore part of nature. . . . We have argued throughout this book that consciousness arises from certain arrangements in the material order of the brain. (pp. 215–219)*

If the word consciousness in Edelman and Tononi's (2000) quote could be replaced with "the mind," or "human mental cognition," the observation might be less problematic, but to state that consciousness arises from the brain is questionable. This essay takes the opposite view, that the human mind, mental cognition, the operational processes of the brain arises from consciousness, that they are not the same thing.

Nor is consciousness necessarily limited to space, or restricted in time, nor exclusive to human primates. Descriptions of mystical and religious experiences handed down by every culture provide strong evidence that there exist modes of consciousness that can be explored beyond normal waking thought; all traditions offer prayer, contemplation, and the ingestion of psychotropic plant substances as doorways to experience beyond space and time (Bellah, 2011). Additional evidence is close at hand—the universal nightly human experience of dream states—which seem not necessarily to be a product of normally experienced time, space, or mental cognition.

It is evident that serious efforts have indeed been made to explore consciousness in order to discover the outlines of an architecture of consciousness through first-hand experience. In support of such efforts, and to counter the epiphenomenalist turn in consciousness studies, this book offers a model of consciousness that builds upon the ideas of the theoretical physicist David Bohm (1980) and the neuroscientist Karl Pribram (2013).

BACKGROUND: The Pribram-Bohm Holoflux Topology

The topological model which is developed here in support of a metaverse cosmology is presented in Figure 1, the "Pribram-Bohm Holoflux Model," where the basic theory is diagrammed as consciousness transforming between nonlocal and local regions of experience and information (Joye, in press).

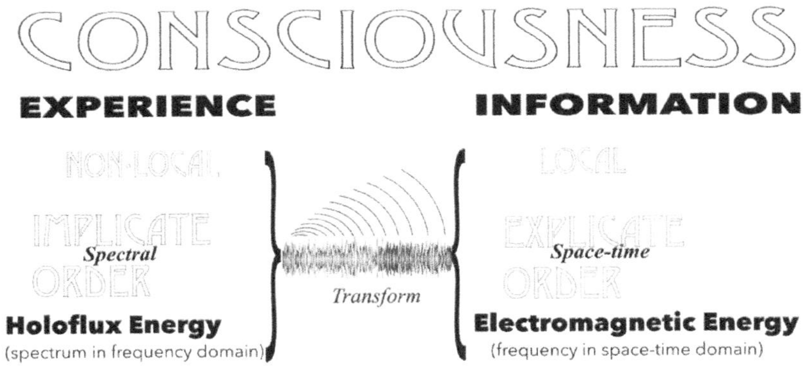

Figure 1. The Pribram-Bohm holoflux model

> *The Pribram-Bohm hypothesis regards consciousness as a cybernetic energy process, a holoflux transforming between two orders of being in "an undivided flowing movement without borders" (Bohm, 1980, p. 172). To the left in the diagram of Figure 1, consciousness is expressed as a spectrum of holoflux energy in Bohm's implicate order. This holoflux energy resonates with electromagnetic energy of the same frequencies to the right in the diagram, in the spacetime region, or explicate order (Bohm, 1980, p. 159).*

Viewed from left to right, the diagram reveals a spectrum of holoflux energy in the transcendental implicate order transforming and translated into "things" and "events" in local spacetime, and conversely, viewing the diagram from right to left, information generated by "things" and "events" interacting throughout spacetime is seen to be transforming (folding) back into the implicate order. The process is described as a continuous cybernetic cycle, perhaps occurring at a regular clock-rate.

Superposition of Consciousness: Cybernetics

Common experience would suggest that all consciousness is consciousness of something, such as the experience of a sound, of an image, of a sensation, of an emotion, of an interior verbal thought. These experiences seem to be superpositioned, they often occur at what seem to be the same perceptual moment. Yet each simultaneous stream of experience remains distinct, somehow integrated with all the others. But what is it that is "looking at" this stream of experiences? It is as if there is some meta-consciousness that is more than the sum of each of these individual "experience streams," per se, but rather some other, more comprehensive level, some panoramic perspective that is able to embrace and comprehend them all, and which has the amazing ability to fine tune its own selected focus upon one or more of these streams of awareness while simultaneously dampening and filtering out the many others.

In signal analysis this stream phenomenon is explained by the superposition principle, formalized in 1822 by the French mathematician Jean-Baptiste Fourier, who developed the mathematics of what is now called "Fourier analysis" during his search for a mathematical relationship between spacetime and frequency (Feynman, Leighton, & Sands, 1964, p. 286). Because signals are more readily superpositioned and manipulated (filtered, amplified, etc.) within the frequency domain than in the time domain, the Fourier transform equations have become primary and ubiquitous mathematical tools in physics and engineering for analyzing, synthesizing, and transmitting signals between two domains:

a "spacetime domain (t_d)," and
a "frequency domain (f_d)."

Much of electrical engineering circuit design is done within the frequency domain, and only subsequently implemented with time domain components, as described here by Francis F. Kuo, chief electrical engineer at the original Bell Telephone Laboratory from which emerged the transistor, the laser, and radio astronomy, information theory, and the Unix operating system. In his textbook on Network Analysis and Synthesis, Kuo (1962) states:

> *We see that in the time domain (i.e., where the independent variable is t) the voltage–current relationships are given in terms of differential equations. On the other hand, in the complex frequency domain, the voltage–current relationships for the elements are expressed in algebraic equations. Algebraic equations are, in most cases, more easily solved than differential equations. Herein lies the raison d'être for describing signals and networks in the frequency domain as well as in the time domain. (p. 13)*

Norbert Wiener (1948) coined the term "cybernetics" from the Greek κυβερνήτης—"steersman, governor, pilot, or rudder" (p. 11)—during his own work at the same Bell Telephone Laboratory as Kuo, and made use of Fourier's transform to model and analyze brain waves in the frequency domain, where he discovered clear evidence of "self-organization of electroencephalograms or brain waves" (p. 181). Using Fourier analysis, an approach which later became of great interest to Bohm, Wiener (1948) was able to detect uniquely narrow frequency ranges, centered within different spatial locations on the cortex, that repeatedly exhibited auto-correlation (p. 191). Regions on the cortex were identified where specific ranges of frequencies were found to coalesce toward intermediate frequencies, seeming both to attract and to strengthen one another, exhibiting resonance or "self tuning" to amplify and consolidate signals into narrowly specific ranges in the frequency domain f_d (p. 198). His research led Wiener to conjecture that the infrared band of electromagnetic flux may be the loci of "self–organizing systems":

> *We thus see that a nonlinear interaction causing the attraction of frequency can generate a self–organizing system, as it does in the case of the brain waves we have discussed. . . . This possibility of self-organization is by no means limited to the very low frequency of these two phenomena. Consider self–organizing systems at the frequency level, say, of infrared light. (p. 202)*

Three years after Wiener's (1948) publication of Cybernetics, David Bohm (1951) stressed the importance of Fourier's equations on the first page of his well-received 646-page textbook, Quantum Theory, where he encouraged a familiarity with Fourier analysis for an ontological understanding of quantum phenomena:

It seems impossible to develop quantum concepts extensively without Fourier analysis. It is, therefore, presupposed that the reader is moderately familiar with Fourier analysis. (p. 1)

For purposes of this book, the basic understanding of Fourier analysis is simply that frequency vibrations manifest within two distinct dimensions or domains: a spacetime domain and a frequency domain. Until recently, physicists have focused exclusively within spacetime to conduct their research, considering only space and time as having any "reality" and considering the ontological reality of the frequency domain, if at all, in the same vague category as the domain of mathematics itself (i.e., in some unspecified transcendent dimension). Whether there might somehow exist a "real" dimension outside of spacetime, or beyond spacetime has generally been beyond the purview of the physical sciences. Yet the experienced reality of a region of consciousness beyond spacetime is supported by the vast body of first–hand reports generated by religious, mystical, or near-death experiences. In an approach to such experiences, William James (1902/2004), the "father of American psychology," writes:

The further limits of our being plunge, it seems to me, into an altogether other dimension of existence from the sensible and merely "understandable" world. Name it the mystical region, or the supernatural region, whichever you choose. So far as our ideal impulses originate in this region (and most of them do originate in it, for we find them possessing us in a way for which we cannot articulately account), we belong to it in a more intimate sense than that in which we belong to the visible world, for we belong in the most intimate sense wherever our ideals belong. (p. 318)

Fourier's transform equations (Figure 2) between the two domains of time (t_d) and frequency (f_d) are more than simply mathematical equations, written down as functions in the abstract symbolic language of calculus (Stein & Shakarchi, 2003, pp. 134–36).

$$f(t) = \int_{-\infty}^{+\infty} X(F) e^{j2\pi Ft} dF \qquad f(F) = \int_{-\infty}^{+\infty} x(t) e^{-j2\pi Ft} dt$$

Fourier integral transform of a continuous frequency function into the time domain (t_d).

Fourier integral transform of a continuous time function into the frequency domain (F_d).

Figure 2. The Fourier transform and inverse transform (Kuo, 1962)

These two expressions indicate that any function in the timespace domain, $f(t)$, can be transformed into and expressed equivalently as an infinite series of frequency spectra functions $X(F)$ in the frequency domain. The transformation is also possible in the opposite direction, such that any arbitrary function in the frequency domain, $f(F)$ can be transformed into and expressed by an infinite series of time functions, $x(t)$. The two domains mirror one another.

Beyond purely mathematical considerations, the equations can be taken as models of an actual cosmic process (i.e., much as Newton's Law model the phenomenon of gravity) and they can be understood as mirroring the cosmos in mathematical terms. The model of consciousness presented in this book proposes that there is indeed an ontological reality to this other region, and that this region is synonymous with Bohm's "implicate order," Pribram's "holonomic frequency domain."

Karl Pribram's Holonomic Mind/Brain Theory: The Frequency Domain

The neurosurgeon Karl Pribram (1971) was one of the first to articulate the idea that the Fourier transform might play a role in brain/mind neurophysics. Pribram (1990) spent decades performing laboratory research to gather experimental data in an effort to solve two problems: (a) to identify the location and mechanism of memory storage (the engram), and (b) to discover the cognitive mechanism behind visual perception. Pribram (2013) arrived at the conclusion that the data revealed evidence of Fourier signal transformations of visual signals from the rods and cones of the eyes, and that these Fourier patterns could be detected in spatial Fourier patterns over wide areas of the brain, as fields within the fine-fibered dendritic networks of the cerebral cortex (p. 82).

In the mid-1960s, Pribram was inspired by reports of the first optical holograms, and the empirical evidence that holograms could store, retrieve, and process vast quantities of information using resonant photons in high frequency beams. Ten years later, Pribram (1971) published Languages of the Brain, in which he detailed his new theory, the holonomic brain/mind theory, based upon evidence of the Fourier transform playing a key role in the mind/brain process. The theory he put forth proposed that the cognitive sensory processes of memory, sight, hearing, and consciousness in general, may all operate holographically, in a transformational process of information-coded-energies flowing back and forth between space–time and the frequency domain via a Fourier transform mechanism.

Pribram's (1971) theory was radical and controversial, challenging two prominent paradigms of modern neurophysical research: (a) the belief that consciousness is an epiphenomenon produced by electrical sparks among synaptic clefts throughout the wiring of neurons the brain, and (b) the belief that somewhere in the physical brain, engrams of memory are stored, and will be eventually found. Pribram (1971) relates a story of a conversation he had at the time, while climbing with colleagues on a hike in Colorado just prior to attending a neuroscience conference in Boulder:

> *We had climbed high into the Rocky Mountains. Coming to rest on a desolate crag, a long meditative silence was suddenly broken by a query from Campbell: "Karl, do you really believe it's a Fourier?" I hesitated, and then replied, "No Fergus, that would be too easy, don't you agree?" Campbell sat silently awhile, then said, "You are right, it's probably not that easy. So what are you going to say tomorrow down there?" I replied, this time without hesitation, "That the transform is a Fourier, of course." Campbell smiled and chortled, "Good for you! So am I." (p. xvii)*

Pribram's (1971) hypothesis was strengthened through a growing appreciation of holography as frequency-superpositioned electromagnetic wave interference (p. 142). Pribram called his approach "the holonomic brain theory," and postulated the importance of the frequency domain in future research:

> *Essentially, the theory reads that the brain at one stage of processing performs its analyses in the frequency domain . . . a solid body of evidence has accumulated that the auditory, somatosensory, motor, and visual systems of the brain do in fact process, at one or several stages, input from the senses in the frequency domain. (Pribram, 1982, p. 29)*

In Pribram's (1990) theory, a pure frequency domain links with the neuronal tissue of the brain through modulating fields of flux within the fine-fibered dendritic webs of the cerebral cortex regions. His paradigm was reinforced at a San Francisco conference during a lecture given by the physicist Geoffrey F. Chew, the head of the UC Berkeley physics department and a former student of Enrico Fermi. Chew presented a conceptual diagram of the Fourier transform process (Figure 3), which perfectly encapsulated what Pribram had by then become familiar with, the Fourier transform (Pribram, 2004a, p. 230). As shown in the figure, the spectral (frequency) domain, located at the left of the diagram, is directly linked to the spacetime domain, depicted at the right, bridged by the Fourier transform, operating at the sub-atomic levels predicted by Planck's constant.

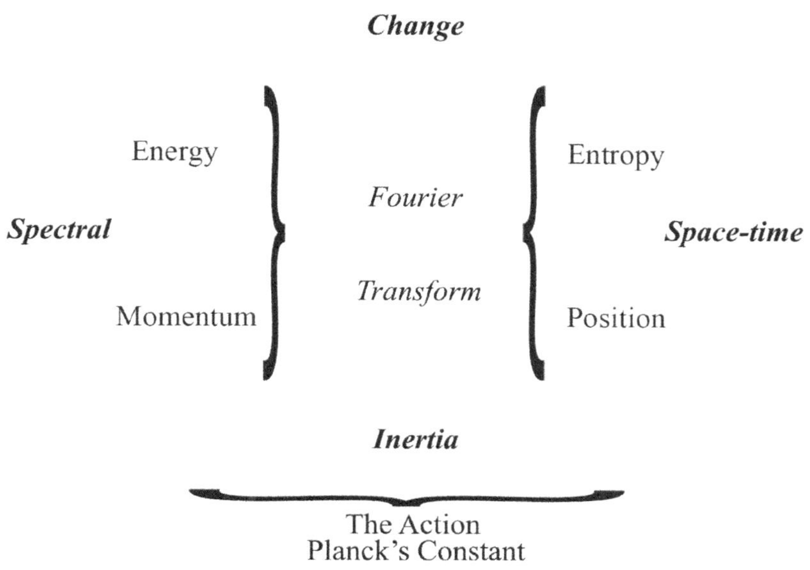

Figure 3. The Dirac Fourier transform diagram. Source: Adapted from Pribram (2004b)

Pribram (2013) asked Chew where he had obtained the diagram, and was told that he had been given the diagram by his colleague at Berkeley, the physicist Henry Stapp, who himself said he had been given it directly from the British theoretical physicist Paul Dirac (1902–1984), one of the original founders of quantum mechanics. Whatever the origin of the figure, Pribram chose to include the diagram in several future books. In "Consciousness Reassessed," Pribram's (2004b) caption to the figure reads, "The Fourier Transform as the Mediator between Spectral and Spacetime" (p. 8).

In the diagram, the spectral domain is shown at the left and spacetime to the right, with the Fourier transform between them. The diagram became foundational to

Pribram's understanding. It presents a two–way Fourier transform, operational at the boundary between the two domains, located at an event horizon termed in the diagram, "The Action: Planck's Constant." It is this process of turbulent transformation at the event–horizon that David Bohm and Basil Hiley (1993) termed holomovement or holoflux (p. 382).

The Limits of Space

From the Edge of the Universe to Planck's Constant

The Pribram–Bohm hypothesis holds that the dimensions of space are finite, and that space exhibits a limited domain in a quantifiable range. This is consistent with the physics of string theory or M–theory, according to Bernard Carr (2007), Professor of Mathematics and Astronomy at the University of London:

> *The Universe may have more than the three dimensions of space that we actually observe, with the extra dimensions being compactified on the Planck scale (the distance of 10^{-35} meters, at which quantum gravity effects become important), so that we do not notice them. . . . In particular, physics has revealed a unity about the Universe which makes it clear that everything is connected in a way which would have seemed inconceivable a few decades ago. The discovery of dark dimensions through particle physics shakes our view of the nature of reality just as profoundly as the discovery of dark energy through cosmology. (p. 10)*

Carr (2007) uses the alchemical image of the ouroboros (Figure 4) to illustrate his GUT theory (Grand Unified Theory) in comparing major scale-dependent structural levels of the physical world: “The significance of the head meeting the tail is that the entire Universe was once compressed to a point of infinite density (or, more strictly, the Planck density)” (p. 13). This archetypal figure implies the interconnectedness of the entire universal process in time and space, presenting a cybernetic feedback loop operational at every scale. Mystics have intuited this ouroboric process symbolized in the images of a snake swallowing its own tail (the image has been found as early as the 14th century BCE in the tomb of Tutankhamun) and it is frequently used to symbolize cybernetic feedback in control and communication theory (Wiener, 1964). While Wiener coined the term cybernetics, communication engineers would more commonly see this as metaphor for the “feedback loop,” used everywhere in electronic circuit design.

Figure 4. Alchemical Ouroboros (Pelekanos, 1478)

Stretching out this circular cosmic ouroboric serpent from head to tail, one can create an axis of scales that encompasses all of space. In Figure 5 such a scale is drawn starting with the currently estimated diameter of the universe itself at 10^{+25} m, and descending logarithmically down to the Planck length limit at 10^{-35} m. The axis thus spans a total range of 10^{+60} (60 jumps by the power of 10). The Pribram–Bohm hypothesis holds that there, at the very bottom of the linear scale (Figure 5), is to be found the transition bounding the explicate order and the implicate order (Joye, in press, p. 261). Here, at the bottom bound of the spatial scale, space reaches its end, according to modern physics; but it also marks the entry point into Bohm's "implicate order," what Pribram terms the "frequency domain" (Bohm, 1980; Pribram, 2013).

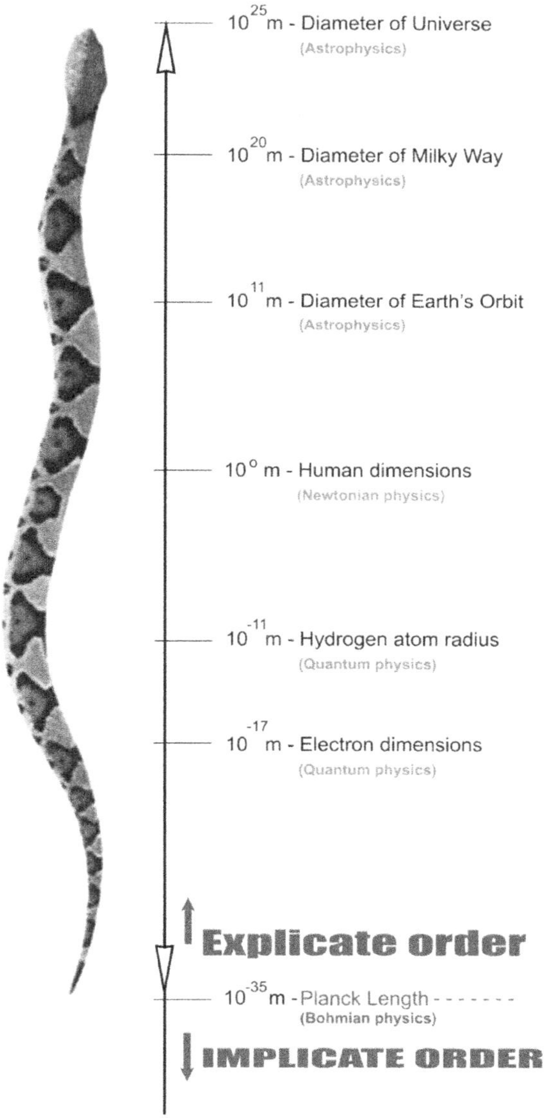

Figure 5. Scales of dimensional space and the explicate/implicate boundary

The implications of this topology are profound. Imagine moving inwardly, from any position in the universe, moving into a spherical bubble, shrinking ever smaller in scale while moving ever closer to the center at the bottom of the spatial scale, following the radial axis inward, ever shrinking downward, and then abruptly reaching the end of the line at the Planck length limit of space, the locus of a spherical shell 10^{-35} meters in diameter, below which space has no meaning. Here a boundary has been reached, an event horizon between space and the implicate order. To understand this, one must realize that the classical Cartesian assumption that space is continuous is wrong; there is indeed a bottom to space, at least according to physics, below which space no longer has meaning. Here there is a discontinuity, as David Bohm and F. David Peat (1987) explain in describing the granularity of space:

> *What of the order between two points in space? The Cartesian order holds that space is continuous. Between any two points, no matter how close they lie, occur an infinite of other points. Between any two neighboring points in this infinity lies another infinity and so on. This notion of continuity is not compatible with the order of quantum theory. . . . Thus the physicist John Wheeler has suggested that, at very short distances, continuous space begins to break up into a foam-like structure. Thus the "order between" two points moves from the order of continuity to an order of a discontinuous foam. (pp. 311–312)*

Pribram's Spectral Density Flux and the Implicate Order

In 1979, Karl Pribram, at that time a Stanford professor, attended a conference in Cordoba, Spain, where he met David Bohm, a professor of theoretical physics at London University (Cazenave, 1984). During the conference, Pribram (2013) soon realized that David Bohm's model of the implicate order and its projection, or extrusion into space–time, could be seen as entirely compatible with his own holonomic mind/brain theory. Thus began 20 years of correspondence and dialog between David Bohm and Karl Pribram, and the two soon became personal friends.

Pribram (2013) saw in Bohm's theories how the frequency domain flux might be seen to unfold into explicate domain waves of encoded information via the Fourier transform, and he appreciated Bohm's description of how information from the explicate may fold back into the implicate in a bi-directional process. Even more intriguing was Bohm's belief that, "the basic relationship of quantum theory and consciousness is that they have the implicate order in common [emphasis added]" (Bohm & Hiley, 1993, pp. 381–382).

Pribram was equally impressed with Bohm's explanation of nonlocality, a major mystery in quantum physics, which Bohm explains as fundamental to the process of folding and unfolding between explicate and implicate orders, allowing for full superpositioned cohesion of frequency information within the implicate order, and even providing a plausible mechanism for

Sheldrake's theories of morphogenetic fields and morphic resonance:

> *The implicate order can be thought of as a ground beyond time, a totality, out of which each moment is projected into the explicate order. For every moment that is projected out into the explicate there would be another movement in which that moment would be injected or "introjected" back into the implicate order. If you have a large number of repetitions of this process, you'll start to build up a fairly constant component to this series of projection and injection. That is, a fixed disposition would become established. The point is that, via this process, past forms would tend to be repeated or replicated in the present, and that is very similar to what Sheldrake calls a morphogenetic field and morphic resonance. Moreover, such a field would not be located anywhere. When it projects back into the totality (the implicate order), since no space and time are relevant there, all things of a similar nature might get connected together or resonate in totality. When the explicate order enfolds into the implicate order, which does not have any space, all places and all times are, we might say, merged, so that what happens in one place will interpenetrate what happens in another place. (Bohm & Weber, 1982, pp. 35–36)*

Bohm's topology is both supported and extended by Pribram's contention, supported by the diagram handed down from Dirac, that the boundary or event horizon between the two domains, where the action occurs, is at the Planck length, precisely where, as Pribram tells us here, spectral density in-formation translates into spacetime ex-formation.

> *Matter can be seen as an "ex-formation," an externalized (extruded, palpable, compacted) form of flux. By contrast, thinking and its communication (minding) are the consequence of an internalized (neg-entropic) forming of flux, its "in-formation." My claim is that the basis function from which both matter and mind are "formed" is flux (measured as spectral density). (Pribram, 2004b, p. 13)*

This flux or spectral density is for Pribram real, in the same sense that spacetime is considered to be real, but this flux is outside of or beyond space–time. It is in this sense that Pribram made the conceptual leap from considering the Fourier transform as simply a tool of mathematical calculation, to a dawning realization that the reality of the transform implies the ontological reality of a domain outside of space–time, a transcendent yet ontologically real domain where energy as flux is "measured as spectral density."

Dirac's original diagram can now be extended to include Bohm's two regions of the whole, the implicate order and the explicate order. Figure 6 depicts this expanded diagram.

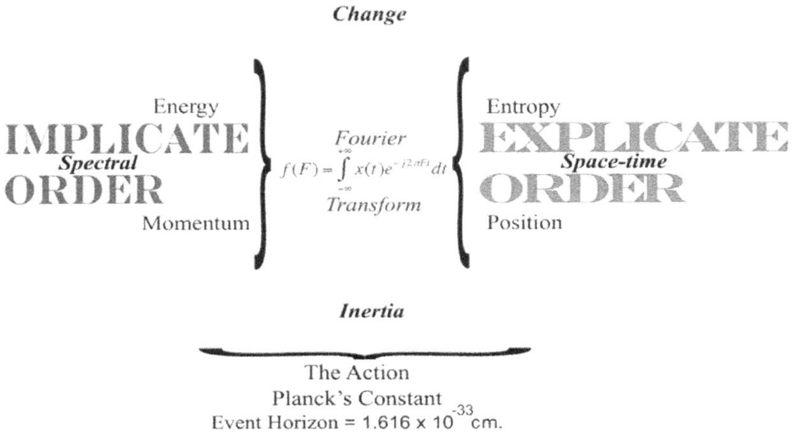

Figure 6. Dirac's Fourier diagram with David Bohm's topology

An anthropomorphic view of the Pribram's diagram can be seen in Figure 7, where an iris-like lens peering out from the implicate order is maintaining a focus upon and/or projecting a holonomic universe within the explicate order of space–time. This mirrors Karl Pribram's (1991) conceptualization of a lens between the two domains, expressed here in Brain and Perception:

> *These two domains characterize the input to and output from a lens that performs a Fourier transform. On one side of the transform lies the spacetime order we ordinarily perceive. On the other side lies a distributed enfolded holographic–like order referred to as the frequency or spectral domain. (p. 70).*

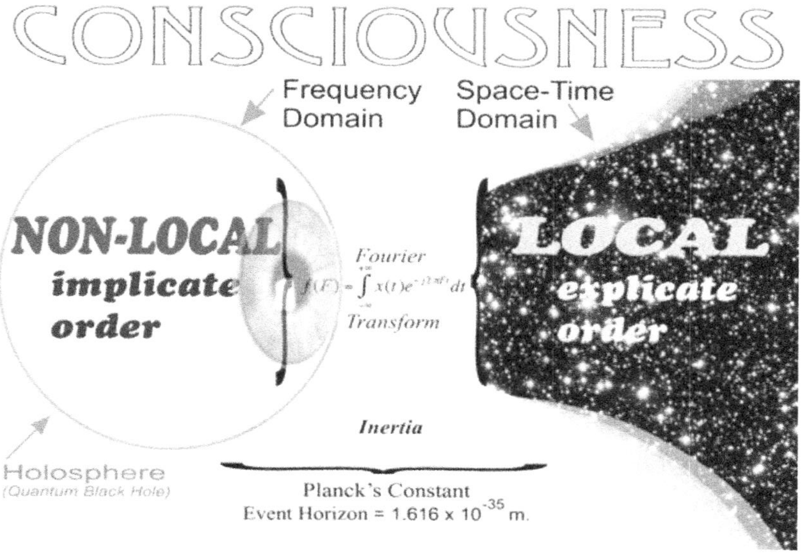

Figure 7. Topology of consciousness in Pribram's diagram

Note that the image of an iris in the diagram appears at the edge of the event horizon of a quantum black hole, or implicate order holosphere. The iris symbolizes consciousness looking out from the implicate order into spacetime via a Fourier transform lensing process. This approach to a topology of consciousness as something that is looking out and seeing itself is supported here by the mathematician G. Spencer-Brown (1972) in Laws of Form:

> *Now the physicist himself, who describes all this, is, in his own account, constructed of it. He is, in short, made of a conglomeration of the very particulars he describes, no more, no less, bound together by and obeying such general laws as he himself has managed to find and record. Thus we cannot escape the fact that the world we know is constructed in order (and thus in such a way as to be able) to see itself. This is indeed amazing. Not*

so much in view of what it sees, although this may appear fantastic enough, but in respect of the fact that it can see at all. . . . But in order to do so, evidently it must first cut itself up into at least one state which sees, and at least one other state which is seen. In this condition it will always partially elude itself. (p. 105)

Cosmology and the Implicate Order

In 1980 Bohm published Wholeness and the Implicate Order, and in a section in which he discusses the cosmology of the implicate order, he puts forth a solution to the problem of "zero-point" energy by regarding the Planck length as the shortest wavelength possible:

If one were to add up the energies of all the "wave-particle" modes of excitation in any region of space, the result would be infinite, because an infinite number of wavelengths is present. However, there is good reason to suppose that one need not keep on adding the energies corresponding to shorter and shorter wavelengths. There may be a certain shortest possible wavelength, so that the total number of modes of excitation, and therefore the energy, would be finite. . . . When this length is estimated it turns out to be about 10^{-35} m. (Bohm, 1980, p. 190)

Bohm (1980) brings up the school of Parmenides and Zeno, which held that all of space is actually a plenum, and he points out that as recently as the last century this same theory was presented in the widely accepted hypothesis of the ether (p. 191). Bohm describes how there is a "holomovement" (p. 151) in this immense sea of "zero-point energy" (p. 190) to be understood as a "undivided flowing movement without borders" (p. 172) and he goes on to state:

> *It is being suggested here, then, that what we perceive through the senses as empty space is actually the plenum, which is the ground for the existence of everything, including ourselves. The things that appear to our senses are derivative forms and their true meaning can be seen only when we consider the plenum, in which they are generated and sustained, and into which they must ultimately vanish. (p. 192)*

In the Pribram–Bohm cosmology then, the interface or boundary between the spacetime explicate domain and the nonlocal, nontemporal implicate domain can be viewed topologically as a holoplenum of holospheres (Figure 8). Here can be found an answer to the "hard problem of consciousness" posed by Chalmers (1995), for it is from within each holosphere that consciousness is "peering out" into and "projecting" the spacetime explicate, and here Bohm (1980) summarizes his cosmological essay by proposing that "consciousness is to be comprehended in terms of the implicate order, along with reality as a whole" (p. 196) and stating unequivocally that "the implicate order is also its primary and immediate actuality" (p. 197).

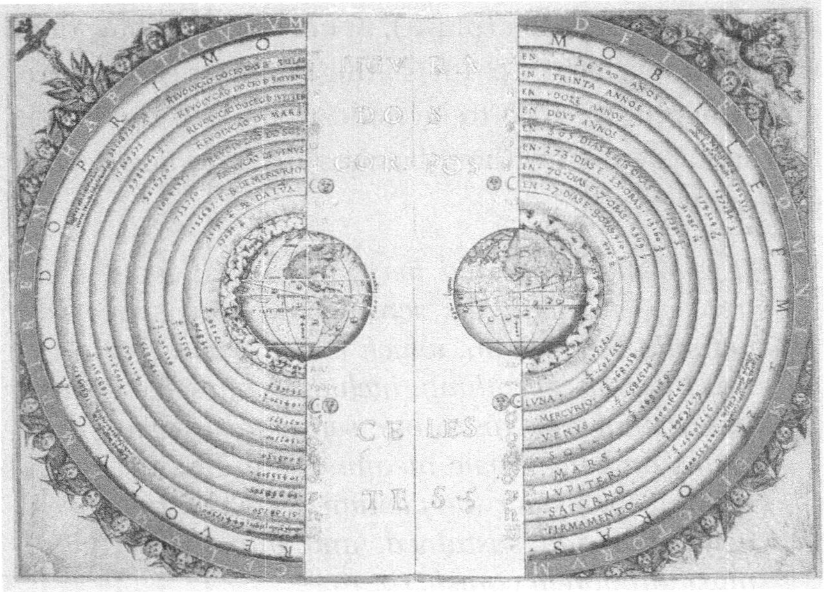

Figure 8. The Ptolemaic Geocentric conception of the Universe. Source: Velho (1568)

In the 3rd century BCE Aristotle (2004) held that the reason an apple falls to the ground is because it seeks its natural place at the center of the universe, and he set forth a geocentric model based upon the following three propositions:

- The Earth is positioned at the center of the universe.
- The Earth is fixed (nonmoving) in relation to the rest of the universe.
- The Earth is special and unique compared to all other heavenly bodies.

Substituting "holosphere" for "Earth" in Aristotle's propositions, each Planck holosphere can be taken as positioned at the center of the universe, each holosphere is fixed (nonmoving) in relation to all other holospheres in the holoplenum, and each holosphere is "special" by virtue of its unique Hilbert space coordinates (Young, 1988).

The Physiology of Consciousness

The first section of this book has been a focus on the theory that the ground of consciousness, consciousness itself, manifests as a continuously transforming flow of energy in a two-way inward/outward flow between two domains whose boundaries are omni-present. This energy flux was described as being encoded with both information and meaning, flowing alternately between the explicate domain of space–time and the nonspatial, nontemporal transcendent domain which David Bohm termed the implicate order, a domain that exists within the center, everywhere, of space–time.

To express the interconnectedness of this energy flux, spanning a wholeness that embraces both space–time phenomena as well as transcendent phenomena, we have selected the term *holoflux*, from the root term holism, defined here as:

> *Holoflux: the fundamental interconnectedness of all phenomena, and within this fundamental interconnectedness not only is each entity, relationship, experience and phenomenon a whole, in its own right, but all wholes are held together by a great unifying force. (Collister, 2010, p. 69).*

The Holoflux Theory of Consciousness

The holoflux theory of consciousness rests upon two fundamental paradigms: Pribram's holonomic brain theory and Bohm's ontological interpretation of quantum theory. These theories have been shown to be congruent, supported and knit together by established principles of electrical communication engineering. As described in Part I of this discussion, the holoflux theory proposed here can be categorized as a cosmological ontological process structured upon these propositions:

- A CENTERED COSMOS: In agreement with the theory of general relativity, the holoflux theory is founded upon the notion that the universe has a center, and that center is everywhere.
- HOLOFLUX: Dark energy of the implicate order, outside of space–time; Karl Pribram's theorized flux within David Bohm's implicate order; it is an energy of sentient awareness that views the outwardly explicate world of space–time from within an implicate order.
- QUANTUM BLACK HOLES: Quantum black holes, Planck holospheres of the Bohmian implicate order; gateway portals into the implicate order, they are bounded by spherical shells each of Planck length diameter 10^{-35} meters.
- THE HOLOPLENUM: An invariant holoplenum of close-packed holospheres, or Planck isospheres, underlying and from which is projected the explicate order; the holoplenum fills a continuum of isospheres everywhere in the physical universe from the center of each position in space–time; the concept had been intuited by Leibniz in his theory of the monads.

- RESONANCE: Holoflux frequency resonance complementarity connects the explicate order (the outside world), with the implicate order (the inside world); holoflux resonance occurs everywhere between the implicate order (frequency-phase flux in the implicate order, f_i) and isospheres of the explicate order in spherical shells. These holo-shells would manifest at corresponding distances, λ_i, from each central Planck holosphere, to maintain the relationship, $f_i \lambda_i$,= c, where c is the speed of light in a vacuum, $3x10^8$ m/sec.
- CYCLIC PROCESS: A flowing movement, an electromagnetic-holoflux movement of energy continually cycles between the implicate order and the explicate order; the flow can be mathematically modeled through Fourier transforms and the Bohm quantum potential wavefunctions.
- A COSMIC CLOCK: Granular units of Planck time (t_P), vibrate at a "clock-cycle" rate of 5.39106×10^{44} cycles per second synchronizing the flow between alternating manifestations of implicate and explicate frequency information; in Cartesian coordinates, this can be pictured as a square wave, a Planck-time clock.
- BOHM'S QUANTUM POTENTIAL FUNCTION: Conceptional computations within the implicate order are "expressed" in space–time (the explicate order) via the effect of Bohm's quantum potential function, Q, a concept pioneered in de Broglie's "pilot wave" paradigm of 1927.

- A FOURIER LENS: The Fourier-type mathematical transform can be seen to act as a lens through which the implicate "views" or "cognizes" the explicate order of space–time; this Fourier lensing is the process by which the implicate order observes, retrieves, and projects information generated within the explicate order.

A diagram of these nine propositions is presented in fig. 9, "Solving the Hard Problem of Consciousness," in which congruent concepts of five contemporary pioneering philosophers of consciousness studies are contextually positioned.

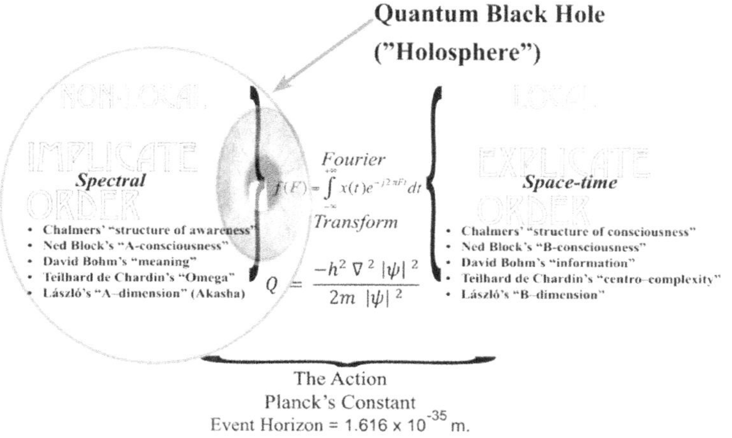

Figure 9. Solving the hard problem of consciousness (Joye, 2016)

In the diagram can be seen five correlated mappings between the nonlocal and the local domains of reality:

- David Chalmers' "structure of awareness" is posited to be synonymous with the implicate order, and his "structure of consciousness" accordingly is shown within the explicate order. (1995).

- Ned Block's "A-consciousness" lies in the implicate order, and his "B-consciousness" in the explicate order. (1995).
- David Bohm's "meaning" lies in the implicate order, and his "information" in the explicate order. (1980).
- Pierre Teilhard de Chardin's "Omega" lies in the implicate order, and his "centro-complexity" in the explicate order. (1956).
- László's "A-dimension" lies in the implicate order, and his "B-dimension" lies in the explicate order. (2014).

The Cosmological Process as a Quantum Potential Plasma Display

As described in The Cosmology of Consciousness, the holonomic process underlying the Bohmian "Whole" operates through means of a holoplenum of quantum black holes, or Planck holospheres, located spatially at the center, everywhere. Surrounding and radiating outward from each central Planck holosphere throughout space are spherical shells of quantum potential, infinitely thin information–encoded shells of energy, each separated from an inner shell by the quantum exclusion radial distance of one Planck length. These isospheric shells of Bohmian quantum potential extend out to the diameter of the universe itself.

Thus can be visualized topologically an almost infinite series of nested isospheric shells, spread out, at discrete quantum radii from their respective central Planck holospheres, each bounding Bohm's nondual implicate order.

The global panoramic intersection of these holospheric shells manifests as the projection of a the holonomic universe into space–time. The cumulative effect of this projection, as regarded by human physicists observing from significantly higher scalar dimensions, is described as "matter." The phenomenon can be understood as a process of projected creation, an omnipresent, on-going holographic extrusion of information from the implicate order into the explicate order, where the various structures of the cosmos (galactic clusters, stars, etc.), the complex unfoldings into space–time, are perceived by the human eye and mind to be three–dimensional, when they are actually holographic projections from the implicate order, from the center outward.

Another way of visualizing the projected illusion of space–time reality from the holoplenum is by expanding upon the metaphor of a flat–panel plasma display (such as the one you may be viewing as you read this).

Consider the human visual threshold for detecting separate images, which lies somewhere between 10 to 12 images per second. The industry standard in the motion picture industry is 24 frames per second. This standard ensures that the presentation of a sequence of projected images will appears to a human viewer as smooth and continuous motion.

By contrast, if the entire universe flashes in and out of existence at a clock-cycle rate limited only by the Planck time constant of 5.3 x 10^{-44} seconds, equivalent to a "frame" rate of almost 10^{44} "frames per second," the cosmos would appear to be smooth and continuous in all respects even to an electron, certainly to any human observer of the cosmos, even at quantum dimensions.

The approximate image resolution of a "holoplenum display" obtained by dividing one inch by the Planck length, yields a maximum resolution of 1.584 x 10^{34} holopixels per inch. At such hyper-fine resolution, even a Higgs boson in the 10^{-17} meter dimensional range would appear to be moving smoothly through space.

Holonomic Information Storage: The Bekenstein Bound

In Wholeness and the Implicate Order, Bohm (1980) articulates and developed a "quantum potential" function which projects the explicate space–time universe out from within an enfolded sub-quantum implicate order (p. 77). Bohm's quantum potential function is congruent with de Broglie's "Pilot wave" theory of 1927, as both are based upon a conviction that there exist "hidden variables" in sub-quantum regions not accessible to observational exploration using current material science technology (and far beyond the capabilities of the CERN Large Hadron Collider).

The de Broglie's pilot wave theory and Bohm's quantum potential are mathematical attempt to map sub-quantum effects issuing from an implicate order in a domain of "hidden variables" far below the observational capabilities of contemporary material science.

Both theories posit a cybernetic processing of information, simultaneously being cycled from the space–time world and enfolded into the nondual frequency domain where the accumulating information is processed nonlocally within the implicate order. Driven then by the implicate order, a pilot wave of quantum potential nudges the configurations in space–time into an altered, slightly new configuration, much as a small tugboat might influence an enormous freighter. If the cosmos operates at its maximum possible clock cycle, as discussed previously, this pilot wave might be seen to operate at the extreme clock cycle rate of the Planck time constant, or 10^{44} Hz.

Somewhere, however, such a cosmic process would require a memory storage repository in space–time. Regarded as a cybernetic process, the sequence of information feedback and action can be metaphorically imaged in the alchemical Ouroboros, the classical symbol for consciousness, depicted as a snake in a circular configuration eating (or chasing) its own tail. This process can be viewed as the cyclic transfer of information coming in (from the tail) and the resulting action (by the head). A cybernetic feedback loop thus needs data, information, as input. Where then might data be accumulated and retrieved in space–time at these most fundamental sub-quantum levels, in a Bohmian holonomic universe consisting topologically of the distributed plenum of Planck holospheres, each surrounded by a series of nested quantum isospheres?

One possibility is to consider the information storage potential of an isosphere encoded with granular "bits" of data. In 1970, Jacob D. Bekenstein, then a graduate student working under John Archibald Wheeler (who himself coined the term "black hole") proposed a novel idea. Beckenstein proposed that there must be an absolute maximum amount of information that can be stored in a finite region of space, and that the Planck constants in quantum theory can be used to determine this limit (Wheeler, 1990). Twenty years later, Bekenstein's theory was extended into what is called the holographic principle by Leonard Susskind (1995), which describes how information within any volume of space can be encoded on a boundary of the region. A description of this configuration is presented here by Wheeler himself (1990), as first related to him by Bekenstein:

> *One unit of entropy (information), one unit of randomness, one unit of disorder, Bekenstein explained to me, must be associated with a bit of area of this order of magnitude (a Planck length square). . . . Thus one unit of entropy is associated with each 1.04 x 10^{-69} square meters of the horizon of a black hole. (p. 222)*

This proposed upper limit to the information that can be contained upon the surface of a specific, finite volume of space has come to be known as the "Bekenstein bound" (Bekenstein, 1973). Symbolically depicted in fig 10 is a topological depiction of the arrangement of information bits, or "qubits," stored on the bounding surface of a spherical volume or isosphere.

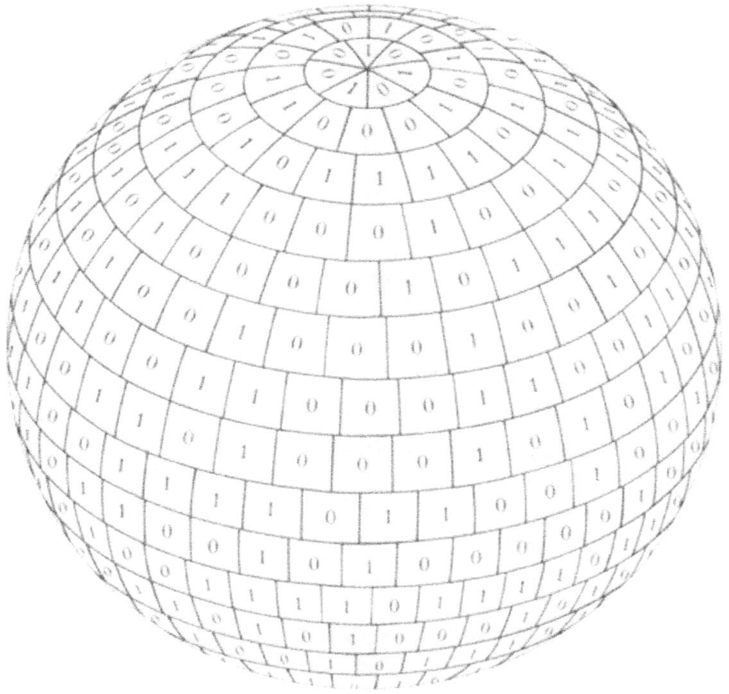

Figure 10. Planck length qubits on the surface of an isosphere. Source: Adapted from Wheeler (1990, p. 220)

This same topological approach to data storage can be applied to human physiology. Using a well-known biological structure as an example, it is possible to calculate the maximum memory storage capacity of a isosphere the size of a single erethrocyte, the ubiquitous red blood cell found throughout the human body. Using Wheeler's approach to determine the number of Bekensteinian equivalent data bits (qubits) on the surface of an isosphere, and using the average diameter of a typical human erethrocyte of 8.1 microns, (or 8.1 x 10^{-5} m), the maximum possible storage capacity on the surface of a single red blood cell can be calculated (Romanes, 1964, p. 137).

To obtain this limiting number of bits, the surface area on a spherical shell 8.1 microns in diameter must be divided by 1.04×10^{-69} square meters (which is the Bekenstein unit of entropy, or approximately the square of the Planck length of 1.616199×10^{-35}). The surface area of this erythrocyte-bisected sphere according to this calculation is $4\pi r^2$ or $4\pi(8.1 \times 10^{-5})^2 = 4\pi(6.561 \times 10^{-9}) = 8.24 \times 10^{-9}$ square meters. Dividing this by the qubit area of 1.04×10^{-69} square meters yields an estimated maximum storage capacity of 8×10^{60} qubits of storage space for potential information encoding. This is an extremely large data storage capacity, considerably larger than, by contrast, the entire projected capability of the National Security Agency's Utah Data Center, which has been designed, when completed, to have a maximum data storage capacity of twelve exabytes or 12×10^{18} bytes (Hill, 2013, para. 7).

Unfolding an Implicate Order into Space–Time: Waveguides and the Topology of Human Consciousness

The basic topology of the Pribram–Bohm hypothesis has now been delineated. The ontological understanding of quantum physics which Bohm sought emerges in this model of a sub–quantum, omni-present holoplenum of Planck holospheres, each enclosing a contiguous transcendent region of non-spatial, nontemporal dimensions termed by Bohm "the implicate order" (Joye, 2016).

What are the implications of this model for human consciousness, cognitively operational at temporal and spatial scales vastly larger than those found at these sub-quantum Planck boundaries?

To answer this we must first complete the Pribram–Bohm cosmological topology of consciousness, and to do this the concept of isospheres, shells within shells, must be considered. Moving outwardly, radially, from the interior bounding event horizon at each central Planck holosphere, can be identified isospheric shells of the implicate order, extruding into space at exact Planck length (quantum) intervals.

This series of concentric shells, each one separated from the next by one Planck length, are isospherical loci of the implicate order (see fig. 11); they extrude into space, and they intersect in space with other shells bounding other Planck holospheres in the space–time holoplenum. It is the cumulative interference effect of the intersection of individual isospheric shells which project images at higher scales, holographically, into three-dimensional space.

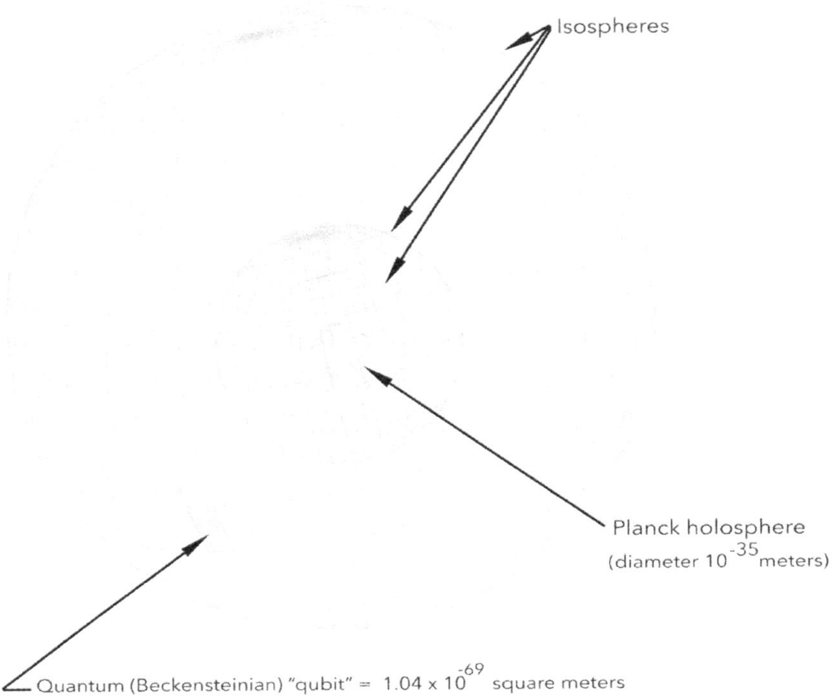

Figure 11. Topology model of isospheres surrounding a Planck holosphere.

Each isospheric shell, as Bekenstein (1973) determined, has a potentially enormous storage capacity in "qubits" of information encoded on the event-horizon bounding each shell, depending only upon the radius of the shell within the range of 10^{-35} m to 10^{27} m. Accessible simultaneously in both the implicate order and the explicate order, such encoded information provides the data to guide evolving forms as they project into the explicate via the pilot wave mathematics of Bohm's (1980) "many-dimensioned quantum potential" (p. 80). As part of this process, in-formation becomes ex-formation as the implicate order unfurls into the explicate. The plasmoidal forms appearing in spacetime as electromagnetic flux energy are mirrored by and resonate within the implicate order as the dark energy of frequency–phase holoflux.

Within this cosmic geometry of the Pribram–Bohm hypothesis can be identified a framework for omnipresent two-way portals, potential bridges between the explicate and the implicate. Here, the deep consciousness of the universe flows in a cyclic, cybernetic, perhaps fractal movement in time and space, moving through an endless, bi-directional process, consciousness involving itself in a dance of transformation.

How then does this topology support human consciousness, thought and perception in space–time? How can electromagnetic frequency plasma in space–time resonate with holoflux plasma in the implicate order? First of all, the energies must be within the same frequency range in order for maximum interactive resonance to occur. Where the frequencies overlap as they superimpose and interpenetrate one other, resonance occurs.

Resonance is a naturally occurring phenomena characteristic of physical objects or plasma fields extended in space–time. The resonance effect is seen when objects or complex signal systems exhibit remarkable frequency sensitivity to particular external frequencies, flowing in, through, and around the system, frequencies which approach the "natural resonant frequency" of the object or signal system; perfect resonance occurs where the input frequency and the natural frequency are identical (Feynman, Leighton, & Sands, 1964, p. 78). This principle of resonance governs all cybernetic feedback loops, and is a key factor in the design of antennas in electromagnetic transmitting and receiving systems.

The goal of antenna design is to construct an antenna that is maximally resonant within a specific narrow frequency range of incoming (external or internal) electromagnetic radiation; when the input frequency and the natural frequency of the antenna coincide or move significantly close to one another, resonance occurs. This simplest and most common antenna design is the dipole antenna, which depends directly on the size (wavelength) of the incoming electromagnetic wave. A dipole antenna is designed to be physically half the size of the incoming wavelength, as stated in a textbook of antenna design:

A fundamental form of an antenna: length is approximately equal to half the transmitting wavelength. It is the unit from which many more complex forms of antennas are constructed. It is known as a dipole antenna. (Blakeslee, 1972, p. 580)

This same half-wavelength effect governs the design of network communication waveguides applied to fiber-optics in the Internet, where the antenna is the fiber channel itself acting as a waveguide. The waveguide is highly efficient for two reasons, first, as its name implies, the waveguide guides the electromagnetic wave within its channel with maximum efficiency, and secondly, it shields the signal in the channel from external electromagnetic waves. The inner diameter of the hollow waveguide is designed to be equal to exactly half the wavelength of the electromagnetic energy signal shielded by and flowing through the waveguide channel.

Waveguides have been used for over a century both commercially and in research to channel and guide vibrating energy of specific limited frequency ranges; the fiber–optic networks hosting the global internet operate on this principle, channeling electromagnetic radiation at fixed laser frequencies (Dorf, 1997). It was discovered late in the 19th century that circular metallic tubes, or hollow metal ducts, similar to A/C ventilation ducts, but much smaller, could be used to channel and guide either sound vibrations in air, or electromagnetic energy in air or vacuum.

Without the waveguide, the vibrational energy field is transmitted in all directions, visualized as magnetic lines or arrows emerging from a point at the center of an expanding sphere. This energy disperses outwardly, the magnetic vectored arrowheads pushing into the inside of an infinitely expanding sphere. A waveguide, however, constrains the magnetic component of the wavefront of vibrating energy to one specific linear direction, in parallel with the center of the waveguide, and thus, conceptually, the confined wave itself loses very little power while it propagates along the central axis of the waveguide, like a stream of water emerging from the pinprick of a large, taut, water balloon (Dorf, 1997).

The most common type of fiber optic cable used in the Internet has a core diameter of 8–10 micrometers and is designed for use in the near infrared (Gowar, 1993, p. 64). The electromagnetic signal wavelength that runs through the global fiber optic network is powered by highly efficient carbon dioxide lasers, and has a wavelength of 10 micrometers. Coincidentally (or not?), the average human blood capillary diameter is also 10 microns, and blood capillaries are at all times full of carbon dioxide.

Dimensional analysis and a cursory examination of human physiology would immediately suggest two candidates for waveguide systems within the human body: (a) within the blood capillary system, and (b) within the microtubule system. The corresponding resonant frequency for electromagnetic waves using such waveguides correspond to wavelengths matching the inner diameter of these structures. For blood system capillaries, this corresponds to radiation with a wavelength of 9.3 to 10.0 microns, the average inner diameter of a capillary. For microtubules, the radiation wavelength would be found in a range of 40 nanometers, the inner diameter of the microtubule waveguides. Figure 12 depicts the location of each of these potential waveguide frequency bands within a wider section of the electromagnetic spectrum.

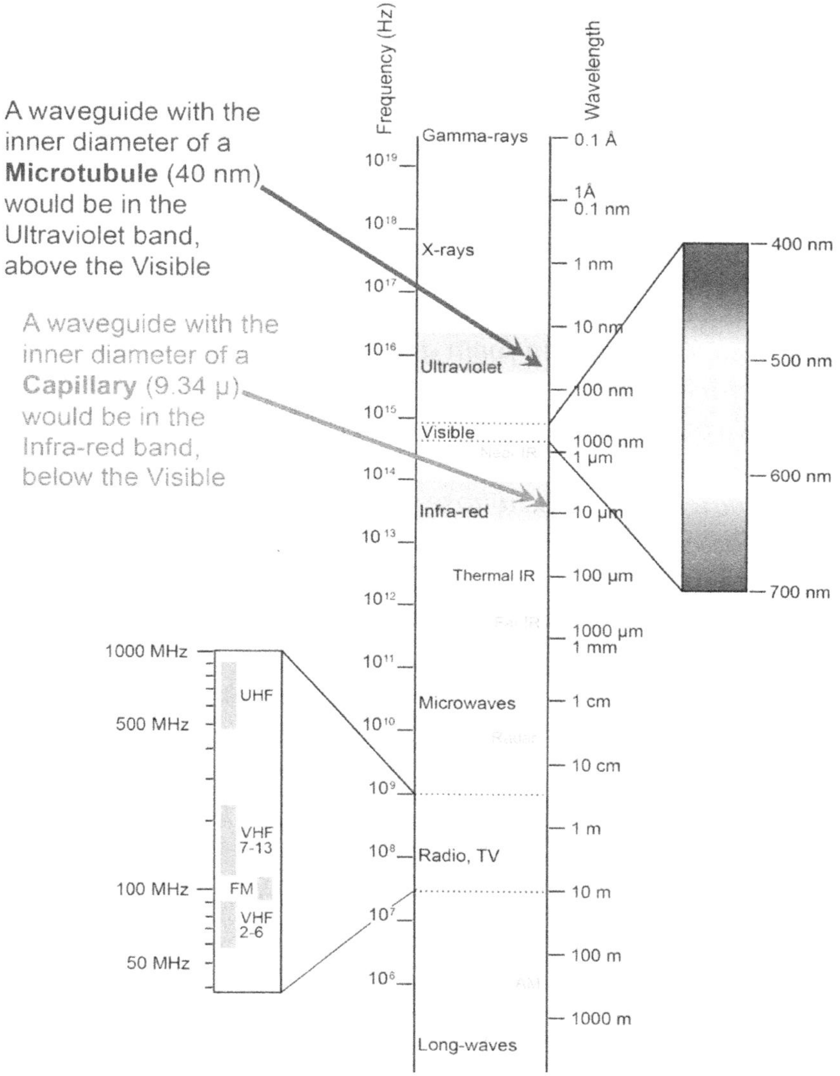

Figure 12. Microtubules and capillaries as waveguides (Joye, 2016)

That our blood might act as an electromagnetic plasma within the capillary system should not be surprising— the opening page in a textbook on plasma physics reads, “It has often been said that 99% of the matter in the universe is in the plasma state" (Chen, 2006, p. 1). Jibu and Yasue (2004) go further by indicating how Quantum Brain Dynamics (QBD) suggests that life itself can be seen to be the equivalent of "the unity of water" in the human body, or as "a single molecule" resonant within the human blood system:

> *All H_2O molecules bound together in a QBD vacuum domain form a single, extensive molecule of water in a macroscopic domain. That is, water throughout the entire region of the cerebral cortex is thought to be composed of many macroscopic water molecules whose sizes are all comparable to the coherence length of the QBD vacuum, that is, about 50 microns. This remarkable feature of water in living matter might provide us with quantum field theoretical support for the idea that life is nothing but the unity of water as a single molecule in living matter. (p. 293)*

A further implication of quantum brain dynamics supports the model of a locus of consciousness beyond that of the nervous system and brain. This runs counter to the widespread assumption that consciousness is simply an epiphenomenon of brain activity. The supporting pattern can be found in the QBD manifestation of a "single, extensive molecule of water," dipolar in geometry like an antenna, generating a complex magnetic field through the circulation of hydrogen ions and carbon dioxide in warm blood plasma.

In such a model, the entire blood system of within the human body can be considered to act as an extensively polarized "super cell" of nonlocal electromagnetic plasma energy, which can then be differentiated from the neuronal brain body of consciousness, itself generated by sequential electrical impulse-driven patterns flowing in the nervous system. Moving charges generate magnetic fields, and ionized human blood flow is no exception: flowing blood plasma results in creation of a magnetic field, and this is in accord the conjecture of QBD (Dorf, 1997, p. 27).

The circulatory system can be seen as a magnetic plasma composed primarily of ionized red blood cells (erythrocytes) and water molecules, flowing together in complex vortices of blood plasma around every cell and through every capillary of the body (McCraty, 2003). Each erythrocyte is a flexible, annular, bio-concave disk shaped like a doughnut (in geometry, a torus), having a thin webbed center where the hole in a pastry doughnut would be located. The typical outside diameter of a red blood cell is approximately 9 microns, close to the infrared wavelength of 9.6 microns generated by the human body (Turgeon, 2012). The adult human body contains approximately 6 grams of iron, of which 60% is stored throughout the 10^{12} erythrocytes, each of which contains approximately 270 million atoms of ionic iron embedded within transparent hemoglobin in a toroidal locus (Romanes, 1964). Thus each erythrocyte, replete with iron ions embedded in hemoglobin, creates in effect, an ionized iron toroid (Wick, Pinggera, & Lehmann, 2000, p. 6).

Recent studies have also discovered neuronal generation of electromagnetic energy in the near infrared region of the spectrum centered around 10 microns. Radiation emission was repeatedly measured emanating from live crab neurons in extremely narrow, discrete spectral bands within the frequency range corresponding to a spectral region from 10.5 to 6.5 microns (A. Fraser & Frey, 1968).

The implications of this model are considerable: there may exist in nature a unique resonant frequency for each individual human being. It is useful here to step through a topological analysis of the possible functions of a human red blood cell, given its geometry, as a locus of consciousness, and the possible use of the erythrocyte as a locus of memory storage at human biological scales. If, as previously conjectured, the red blood cell has an ideal diameter to resonate electromagnetic radiation in its ferrite–embedded ring at the human infrared wavelength of 10 microns, then it is reasonable to ask if this configuration could accommodate a single unique isospheric frequency (wavelength) for each of the currently 7 billion living humans on the planet. In other words, does this geometry allow for the possibility of each unique human being to also have a single unique frequency within the infrared electromagnetic radiation that resonates within the human cardiovascular waveguide system? Figure 13 outlines the topological feasibility of this approach.

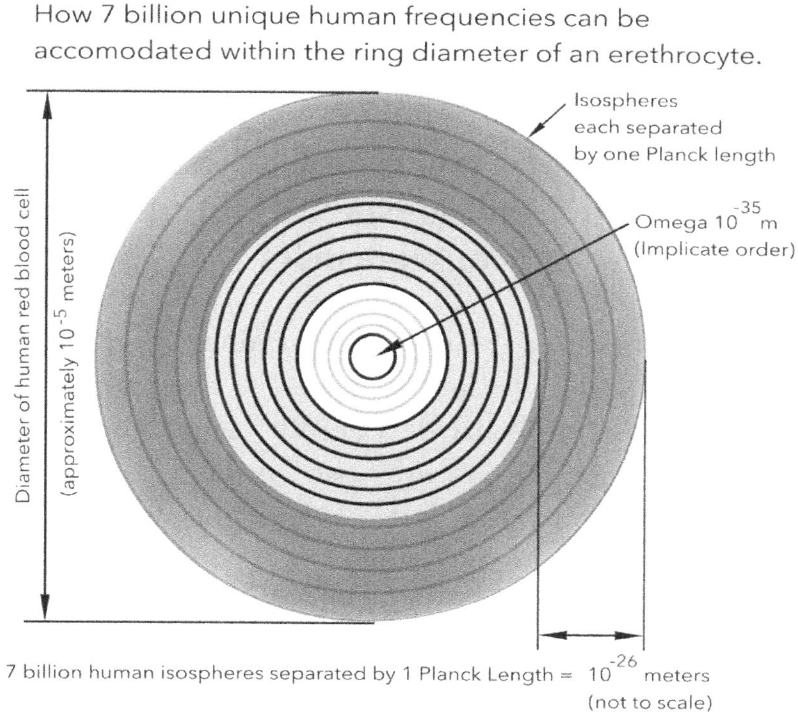

Figure 13. Isospheric capacity of a single erethrocyte (Joye, 2016)

Assuming each unique frequency would match its radially unique isosphere, separated by only one Planck length, Figure 5 suggests how 7 billion unique isospheres, each of quantum discrete frequency, might be nested within the geometry of a typical human red blood cell (in the image, a multiple of 7 billion times the Planck length of approximately 10^{-35} m results in an estimated shell thickness of 10^{-26} m). This supports the feasibility that each living human being might have a unique holospheric frequency, detectable by other human blood cells via the implicate order about which each is centered, and thus provides a possible mechanism for communication, via the mechanisms of resonance, nonlocality, and superposition in the frequency domain of the implicate order.

Two Primary Modes of Consciousness: Explicate Consciousness and Implicate Consciousness

According to the two highly trained, professional, methodologically introspective explorers of consciousness, Teilhard de Chardin (1937) and Rudolf Steiner (1923), there are two primary modes of consciousness. In the last page of his essay, "The Activation of Human Energy," Teilhard de Chardin (1953a) states, "there are two different energies one axial, increasing, and irreversible, and the other peripheral or tangential, constant, and reversible: and these two energies are linked together in 'arrangement'" (p. 393). Likewise, during his discussion of a human physiology of consciousness, Rudolf Steiner (1951) provides a description, replete with diagrams, of how "nerve–activity" (p. 42) must be seen to be at right angles to the alignment of the "blood–activity" (p. 42) in consciousness. In the Pribram–Bohm hypothesis these two primary modes of consciousness are characterized rather generally as implicate consciousness and explicate consciousness.

Teilhard de Chardin (1953a) goes further and links the axial component of consciousness to the process of crossing "a certain critical point of centration" (p. 106), thus thermodynamically heating up consciousness while moving radially inward to resonate with ever smaller radial wavelengths. Teilhard also occasionally refers to radial consciousness as "soul" and "spirit" in his essays (perhaps to satisfy superiors in his religious order), and he also describes this radial component of energy as "a new dimensional zone" (Teilhard de Chardin, 1953b, p. 29) that brings with it "new properties" (p. 29). Teilhard (1953a) describes how increasing centration along the radial component leads to increasing states of complexity–consciousness which result in "being mentally ultra–humanized by self–compression" (p. 344).

Congruently, Steiner (1951) and Teilhard (1953a) both agree that these modes of awareness can be seen to operate orthogonally to one another, each in their own domain, and thus they both refer to consciousness as operating in two modes: a tangential consciousness and an axial consciousness. It is of interest to note that Steiner (1951) says that one of these primary modes, the radial mode of consciousness, operates within the blood system (p. 43).

The Pribram–Bohm hypothesis posits that human consciousness, what is normally experienced as awareness by humans who are not sleeping, nor in a coma, is tangential consciousness, as it manifests, in accordance with Pribram's research, in the resonance of electromagnetic plasma fields swirling in space–time within the felt–like dendritic neuronal regions of the cerebral cortex. This is the "consciousness" examined in experiments conducted by Benjamin Libet (2004) whose research revealed that human awareness of an "intention to act" (p. 86) lagged behind the actual EEG measured "readiness potential" (p. 87) by an average of 200 milliseconds.

This evidence supported the currently widely held view among neurophysiologists that consciousness is an epiphenomenon of neuronal activity. This epiphenomenal view is correct only if one mode of consciousness is being considered, what Teilhard and Steiner would call the tangential mode of human cognitive mental consciousness, which interfaces most generally with the neuronal synaptic activity of the nervous system throughout the human body.

The actual situation is more complex in the Pribram–Bohm holoflux model, in which two modes of consciousness function in a cybernetic process depicted visually in fig. 9, where the Teilhard/Steiner axial consciousness is seen to "peer out from" the transcendent implicate order on the left into the explicate space–time order on the right while simultaneously projecting the explicate configuration itself. The tangential mode of consciousness in the spacetime explicate domain lags, temporally, behind the axial, due to its operational requirement of comparing slices of mnemonic information storage in the time stream. Given the various operating cycles of physiological systems with the human body, this lag is measured to be an average of 200 milliseconds after cognition by the more immediate, axial, implicate consciousness.

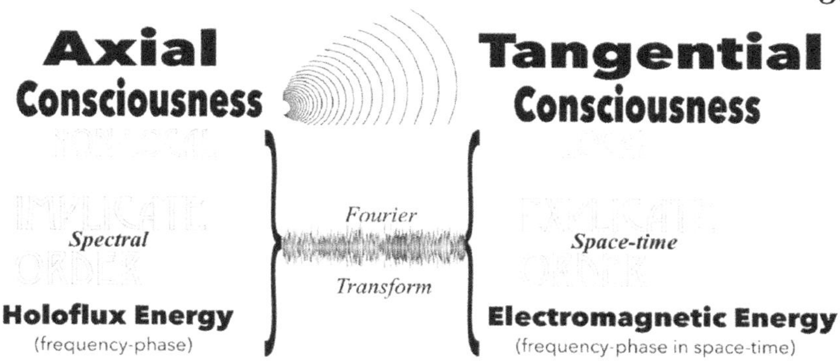

Figure 14. Axial consciousness and tangential consciousness (Joye, 2016)

This timing is in accordance with Libet's (2004) experimental data, and Bohm uses the phenomenon to draw a distinction between thought and a more primary mode of consciousness that is "before thought," the mode of consciousness that dwells in the present moment:

> *To dwell in the present moment requires not dwelling in thought, because thought takes duration and is a slow process. We start to find a need, not to deny thought, but to find the part of the experience with is before thought (Bohm, 1985, p. 128).*

SOLUTIONS AND RECOMMENDATIONS

The topological model set forth in this essay provides a feasible solution to the Chalmers (1995) "hard problem of consciousness." If consciousness is considered to be manifesting as energy flow in electromagnetic-frequency fields, then one should be able to determine experimentally the location of high information bandwidth channels within human physiology. Such channels must provide the data network infrastructure through which electromagnetic information interchange guides the growth process, effects repair, and catalyzes evolutionary mutation. The widespread assumption of contemporary neuroscience has been that consciousness emerges from neuronal activity in the human brain only as an epiphenomenon (Dennett, 1992, p. 406). Accordingly, there are regions within which a search would be recommended, outside of the domain of neuronal linkages and synaptic potential dynamics.

DIRECTIONS FOR FUTURE RESEARCH

Dimensional analysis indicates that good candidate ranges for testing the electromagnetic field component of consciousness can be found in the near infrared spectrum. This region lies just below the threshold of the visible spectrum that is picked up by eye cone structures (Oyster, 1999). Human core body temperatures, ranging from 36.3 C to 37.5 C on a diurnal cycle, indicate that, applying Wien's Law, a search should be conducted within the infrared spectrum in a bandwidth between 9.36425 microns and 9.32808 microns, near the center of the infrared portion of the radiant spectrum (Becker, 1990).

Converting from wavelength, this range is equivalent to a frequency range of 30.3 gigahertz to 32.0 gigahertz, an enormously wide band compared to, for example, the FM radio frequency band, ranging from 87.0 to 88.1 MHz (Blakeslee, 1972). Assuming a bandwidth of 200 KHz (typically used for a single FM station), over 8000 equivalent FM radio stations could be broadcast within the human infrared radiation band, with no overlapping interference.

One approach in the search for an infrared component of consciousness would be to monitor the dynamics of an infrared spectrum emanating from within the human body in an attempt to detect information–carrying photons escaping the body as modulated infrared radiation. It is possible to detect wavelengths down to 5 microns (the far infrared region) using Fourier transform infrared spectroscopy and to record non–invasively site–specific emissions of infrared radiation issuing from within human organs (Griffiths & de Haseth, 2007).

> *Interestingly, because biologic materials are transparent to light in the near-infrared region of the light spectrum, transmission of photons through organs is possible. (Cohn, 2007, p. 323)*

The next step would be to demodulate (decode) these photon packet streams, the difficulty being that even in human communication technology there currently exist dozens of modulation techniques (American Radio Relay League, 2016, p. 42).

An alternate approach would be to search for infrared energy signals flowing as patterns within potential physiological waveguide channels located within the human body. The ubiquitous blood capillary system, for example, with typical inner diameters of 10 microns, is a likely candidate to act as an infrared waveguide. Capillaries provide a ready–made network infrastructure within which the flow and resonance of a modulated infrared energy plasma might be discovered. Figure 15 indicates the location of feasible electromagnetic bands of consciousness correlated with existing structural systems of the human body.

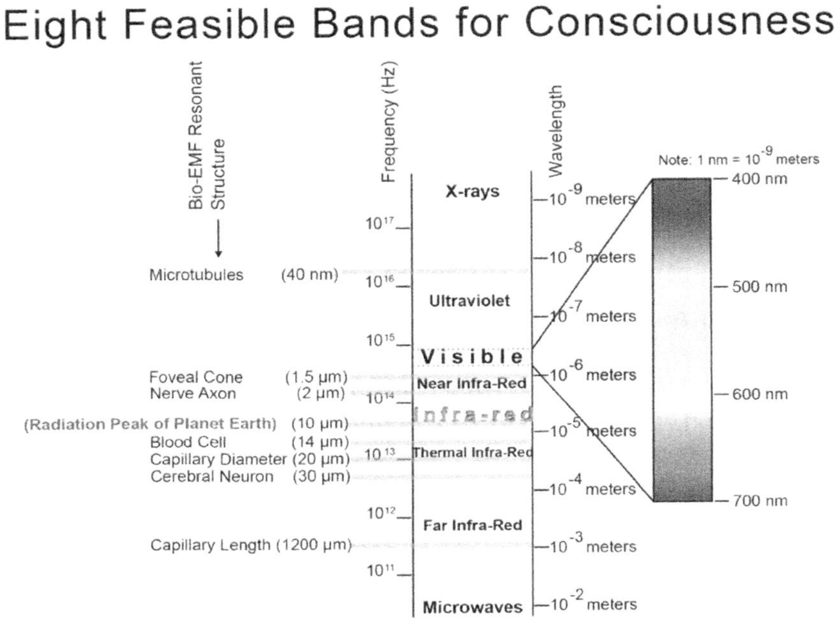

Figure 15. Eight feasible electromagnetic radiation bands of consciousness (Joye, 2016)

As indicated in Figure 15, potential waveguide structures can be found to exist within the human body ranging in diameter (recall that diameter correlates to a dipole wavelength resonant frequency) from ultraviolet radiation in microtubules at a wavelength of approximately 10^{-8} m to erethrocyte radiation in capillaries located in the 10^{-3} m wavelength region.

CONCLUSIONS

Finally, it is reasonable to speculate on practical implications for this topology on future developments. In considering future technologies, the identification of an energy field manifesting information characteristics in multiple bandwidths of radiant energy associated with biological systems would support the feasibility that, in principle, major components of consciousness might eventually be stored, maintained, and manipulated using non-biological systems (e.g., fiber optics and silicon, perhaps) instead of wetware (biological tissue). An intriguing implication would be that consciousness might be uploaded from wetware into such a hardware environment.

Current mainstream research, however, is almost exclusively directed at neuronal activity and brain wiring, under the tacit assumption that consciousness is somehow an accident of, or at best, an epiphenomenon of neuronal activity in the brain. It is time to open our eyes to wider horizons.

REFERENCES

American Radio Relay League. (2015). The ARRL 2016 handbook for radio communications software (93rd ed.). Newington, CT: Author.

Bargh, J. A., & Ferguson, M. J. (2000). Beyond behaviorism: On the automaticity of higher mental processes. Psychological Bulletin, 126(6), 925–945.

Becker, R. O. (1990). Cross currents: The perils of electropollution, the promise of electromedicine. New York, NY: Jeremy P. Tarcher/Penguin Books.

Bekenstein, J. D. (1973). Black holes and entropy. Physical Review, 7(8), 2333–2246.

Blakeslee, D. (Ed.). (1972). The radio amateur's handbook. Newington, CT: Amateur Radio Relay League.

Block, N. (1995). On a confusion about a function of consciousness. Behavioral and Brain Science, 18, 227–287.

Bohm, D. (1951). Quantum theory. New York, NY: Prentice-Hall.

Bohm, D. (1980). Wholeness and the implicate order. London, UK: Routledge & Kegan Paul.

Bohm, D. (1985). Unfolding meaning: A weekend of dialogue with David Bohm (D. Factor, Ed.). Abingdon, UK: Routledge.

Bohm, D., & Hiley, B. J. (1993). The undivided universe: An ontological interpretation of quantum theory. Abingdon, UK: Routledge.

Bohm, D., & Peat, F. D. (1987). Science, order, and creativity. New York, NY: Bantam Books.

Bohm, D., & Weber, R. (1982). Nature as creativity. ReVision, 5(2), 35–40.

Carr, B. (Ed.). (2007). *Universe or multiverse?* Cambridge, UK: Cambridge University Press.

Cazenave, M. (Ed.). (1984). Science and consciousness—Two views of the universe: Edited proceedings of the France-Culture and Radio-France Colloquium, Cordoba, Spain (A. Hall & E. Callender, Trans.). Oxford, UK: Pergamon Press.

Chalmers, D. J. (1995). Facing up to the problem of consciousness. Journal of Consciousness Studies, 2(3), 200–219.

Chalmers, D. J. (2010). The character of consciousness. New York, NY: Oxford University Press.

Chen, F. (2006). Introduction to plasma physics and controlled fusion: Vol. 1. Plasma physics (2nd ed.). New York, NY: Springer.

Cohn, S. (2007). Near-infrared spectroscopy: Potential clinical benefits in surgery. Journal of the American College of Surgeons, 205(2), 322–32. doi:10.1016/j.jamcollsurg.2007.02.024

Collister, R. (2010). A Journey in search of wholeness and meaning, New York: Peter Lang..

Dennett, D. (1992). *Consciousness explained.* New York, NY: Back Bay Books.

Dewey, B. (1985). *The theory of laminated spacetime.* Inverness, CA: Bartholomew Books.

Dorf, R. (Ed.). (1997). *The electrical engineering handbook* (2nd ed.). Boca Raton, FL: CRC Press.

Edelman, G. M., & Tononi, G. (2000). *A universe of consciousness: How matter becomes imagination.* New York, NY: Basic Books.

Feynman, R., Leighton, R., & Sands, M. (1964). The Feynman lectures on physics, Vol. 1. Reading, MA: Addison-Wesley.

Fraser, A., & Frey, O. (1968). Electromagnetic emission at micron wavelengths from active nerves. Biophysical Journal, 8(6), 731–734. doi:10.1016/S0006-3495(68)86517-8

Fraser, C. G. (2006). The cosmos: A historical perspective. Westport, CT: Greenwood Press.

Gowar, J. (1993). Optical communication systems (2d ed.). Hempstead, UK: Prentice-Hall.

Griffiths, P. R., & de Haseth, J. A. (2007). Fourier transform infrared spectrometry (2nd ed.). Hoboken, NJ: Wiley Interscience.

Hardie, W.F.R. (1968). Aristotle's ethical theory. Oxford: Oxford University Press.

Hawking, S. (1988). A brief history of time. New York: Bantam Books.

Hill, K. (2013, July 24). Blueprints of NSA's ridiculously expensive data center in Utah suggest it holds less info than thought. Forbes. Retrieved from http://www.forbes.com/sites/kashmirhill/2013/07/24/blueprints-of-nsa-data-center-in-utah-suggest-its-storage-capacity-is-less-impressive-than-thought/#ae48dc91c85a

James, W. (2004/1902). The varieties of religious experience (reprint ed.). Helena, MT: Kessinger.

Jibu, M., & Yasue, K. (1995). *Quantum brain dynamics and consciousness: An introduction.* Philadelphia, PA: John Benjamins.

Jibu, M., & Yasue, K. (2004). Quantum brain dynamics and quantum field theory. In G. Globus, K. Pribram, & G. Vitiello (Eds.), Brain and being: At the boundary between science, philosophy, language and arts (pp. 267–290). Philadelphia, PA: John Benjamins.

Joye, S. (2016). The Pribram–Bohm holoflux theory of consciousness: An integral interpretation of the theories of

Karl Pribram, David Bohm, and Pierre Teilhard de Chardin (Doctoral dissertation). Retrieved from ProQuest Dissertations and Theses database (Dissertation No. 10117892).

Kuo, F. (1962). Network analysis and synthesis. New York, NY: Wiley.

László, E. (2014). The self-actualizing cosmos: The Akasha revolution in science and human consciousness. Rochester, VT: Inner Traditions.

Levy, N. (2014). Consciousness and moral responsibility. New York, NY: Oxford University Press.

Libet, B. (2004). Mind time—The temporal factor in consciousness. Cambridge, MA: Harvard University Press.

McCraty, R. (2003). The energetic heart: Bioelectromagnetic interactions within and between people. Boulder Creek, CA: Institute of HeartMath.

National Institute of Standards and Technology. (n.d.). Fundamental physical constants: Planck time. Retrieved from http://physics.nist.gov/cgi-bin/cuu/Value?plkt

Norton, D. (Ed.). (1993). The Cambridge companion to Hume. Cambridge: Cambridge University.

Oyster, C. W. (1999). The human eye: Structure and function. Sunderland, MA: Sinauer Associates.

Peat, F.D. (1997). Infinite potential: The life and times of David Bohm. Reading, MA: Addison-Wesley

Pelekanos. (1478). Ouroboros. Reprinted under the terms of a Creative Commons Attribution ShareAlike 3.0 Unported license. Retrieved from Wikimedia Commons. https://commons.wikimedia.org/wiki/File:Serpiente_alquimica.jpg

Pribram, K. (1971). Languages of the brain: Experimental paradoxes and principles in neuropsychology. Englewood Cliffs, NJ: Prentice-Hall.

Pribram, K. (1982). What the fuss is all about. In K. Wilber (Ed.), *The holographic paradigm and other paradoxes* (pp. 27–34). Boulder, CO: Shambhala.

Pribram, K. (1990). Prolegomenon for a holonomic brain theory. In H. Haken & M. Stadler (Eds.), *Synergetics of cognition: Proceedings of the international symposium at Schloss Elmau, Bavaria, June 4–8, 1989* (pp. 150–184). Berlin, Germany: Springer-Verlag.

Pribram, K. (1991). Brain and perception: Holonomy and structure in figural processing. Hillsdale, NJ: Lawrence Erlbaum Associates.

Pribram, K. (2004a). Brain and mathematics. In Brain and being: At the boundary between science, philosophy, language and arts. In G. Globus, K. Pribram, G. Vitiello (Eds.), Brain and being: At the boundary between science, philosophy, language and arts (pp. 215–240). Philadelphia, PA: John Benjamins.

Pribram, K. (2004b). Consciousness reassessed. Mind and Matter, 2(1), 7–35.

Pribram, K. (2013). The form within: My point of view. Westport, CT: Prospecta Press.

Romanes, G. (Ed.). (1964). *Cunningham's textbook of anatomy (10th ed.)*. New York, NY: Oxford University Press.

Ross, W.D. (1930). The right and the good. Oxford, UK: Oxford University Press.

Sheldrake, R. (1981). A new science of life: The hypothesis of morphic resonance. Rochester, VT: Park Street Press.

Smolin, L. (2012). Atoms of space and time. In the Editors of Scientific American (Eds.), A question of time: The ultimate paradox [Kindle DX version]. New York, NY: Scientific American. Available from https://www.amazon.com/Question-Time-Ultimate-Paradox-ebook/dp/B00AF692ZS

Spencer–Brown, G. (1972). Laws of form. New York, NY: Julian Press.

Stein, E., & Shakarchi, R. (2003). Fourier analysis: An introduction. Princeton, NJ: Princeton University Press.

Steiner, R. (1951). An occult physiology: Eight lectures by Rudolf Steiner, given in Prague, 20th to 28th March, 1911 (2nd ed.). London, UK: Rudolf Steiner Publishing.

Susskind, L. (2008). The black hole war: My battle with Stephen Hawking to make the world safe for quantum mechanics. New York: Little, Brown and Company.

Teilhard de Chardin, P. (1969/1937). The phenomenon of spirituality. In *Human Energy* (J. M. Cohen, Trans., pp. 93–112). New York: Harcourt Brace Jovanovitch.

Teilhard de Chardin, P. (1953a). The activation of human energy. In Activation of Energy (R. Hague, Trans., pp. 359–393). London, UK: William Collins Sons.

Teilhard de Chardin, P. (1953b). The atomism of spirit. In Activation of Energy (R. Hague, Trans., pp. 21–57). London: William Collins Sons.

Teilhard de Chardin, P. (1956) The nature of the point Omega. In The Appearance of Man, translated by J.M. Cohen, 271–73. New York: Harper & Row.

Turgeon, M. L. (2012). Clinical hematology: Theory and procedures (5th ed.). Baltimore, MD: Lippincott Williams & Wilkins.

Velho, B. (1568). The Ptolemaic geocentric model of the Universe. Reprinted under the terms of a Creative Commons Attribution ShareAlike 3.0 Unported license. Retrieved from Wikimedia Commons. https://en.wikipedia.org/wiki/File:Bartolomeu_Velho_1568.jpg

Wheeler, J. (1990). A journey into gravity and spacetime. New York, NY: Scientific American.

Wheeler, J. (1998). Geons, black holes, and quantum foam: A life in physics. New York, NY: W. W. Norton.

Whitehead, A.N. (1978). Process and reality: An essay in cosmology. Gifford Lectures, 1927–28. New York: Simon & Schuster.

Wick, M., Pinggera, W., & Lehmann, P. (2011). Clinical aspects and laboratory iron metabolism and anemias (6th ed.). New York, NY: Springer–Verlag.

Wiener, N. (1948). *Cybernetics or control and communication in the animal and the machine.* Cambridge, MA: MIT Press.

Wiener, N. (1964). *God and Golem, Inc.: A comment on certain points where cybernetics impinges on religion.* Cambridge, MA: MIT Press.

Young, N. (1988). An introduction to Hilbert space. Cambridge, UK: Cambridge University Press.

Made in the USA
Monee, IL
04 November 2026

67382366R00046